I0100945

FACT CHECKERS

ANIMAL

MYTHS

&

40+
Amazing
Animal
Facts

MISCONCEPTIONS

By Carrie Rodell and Kizzi Roberts

LearningSpark EDUCATIONAL PUBLISHING

HAVE YOU EVER WONDERED...

DO ELEPHANT GRAVEYARDS EXIST?

DO ELEPHANTS LOVE TO EAT PEANUTS?

DOES A SNAKE USE ITS TONGUE TO SMELL?

FACT FILE

This book is full of myths and misconceptions about animals. Lucky for you, it's also full of wondrous facts. **Fact Checkers** will set the record straight once and for all!

▶ FACT CHECKERS SAY...

KEEP READING!

DO SHARKS DIE IF THEY STOP SWIMMING?

FACT FILE

All spiders have eight legs.

All spiders are arachnids, but not all arachnids are spiders.

Arachnida is a class of animals that includes scorpions, ticks, and more eight-legged animals.

SLEEP AND SPIDERS

Spiders are outdoor creatures, but sometimes they find their way into houses. What happens when a curious house spider crosses paths with a sleeping person? Do we really swallow spiders if we sleep with our mouths open?

DO ALL SPIDERS MAKE WEBS?

All spiders produce silk, but not all spiders make webs. Some spiders make webs to catch their prey. Others, like this wolf spider, don't spin webs to catch prey. Wolf spiders hunt their prey.

FACT CHECKERS SAY...

NOPE!

Many people believe humans swallow an average of eight spiders per year while sleeping. However, there is no evidence to support this. In fact, most spiders prefer being hidden and try to avoid humans.

ELEPHANT GRAVEYARDS

Sometimes large numbers of elephant bones are found together.
Do elephants go to a particular place when they are near death?

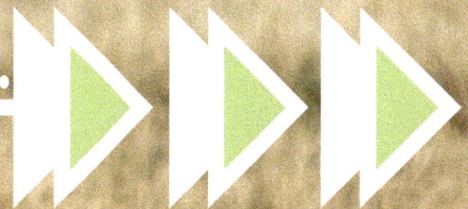

FACT CHECKERS SAY...

NOT ACCURATE

Elephants don't go to a designated place to die. When old elephants can no longer keep up with the other elephants, they leave the herd. These old elephants stay near water and food sources. Over time, old elephants may end up at the same food and water source. The bones of many elephants might be found in one place, but not because the area is a graveyard.

FACT CHECKER REPORT

DO ELEPHANTS LOVE PEANUTS?

Elephants are herbivores, which means they eat a variety of fruits, vegetables, grasses, and trees both in captivity and in the wild. However, peanuts are not part of their normal diet.

FACT FILE

An elephant has four molars plus two tusks (or tushes for female Asian elephants).

An elephant goes through six sets of molars during its life.

After losing its last set of molars, an old elephant can only eat soft vegetation.

DO ELEPHANTS DRINK WATER THROUGH THEIR TRUNKS?

An elephant uses its trunk to drink water, but it doesn't use its trunk like a straw. An elephant sucks water into its trunk and then sprays the water into its mouth to drink.

ARE ELEPHANTS SCARED OF MICE?

Elephants can be startled by objects that move fast across their field of vision. However, there is no evidence they are scared of mice in particular. An elephant may be startled by a mouse or any other small creature that surprises it by moving fast and unexpectedly.

DRAGONFLY DAY

Does a dragonfly only live for 24 hours?

FACT FILE

Dragonflies can hover in the air like a helicopter. This skill comes in handy when looking for food.

NO!

Dragonflies live several years in their nymph stage. Most adult dragonflies live for about a month. Some kinds of dragonflies may even live several months.

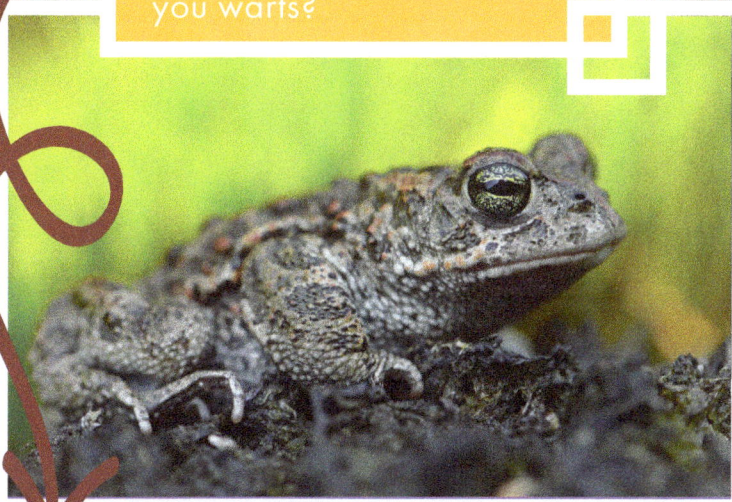

COLOR-CHANGING CHAMELEONS
Can chamelons change color to match any background?

WARTY WONDER
Does touching a toad give you warts?

NOPE!
Thanks to cartoons, chameleons are believed to match different backgrounds exactly. In real life, however, chameleons can only change color within a limited range of colors.

NOT TRUE!
Toads don't give warts to humans. However, touching a toad could irritate your skin.

FACT CHECKER REPORT
Chameleons often change colors not to blend in with their environment, but to stand out. For example, males might display bright colors to attract a mate or when fighting another male.

STINGY STINGER

The buzzing of an angry bee is frightening!
A bee sting hurts, but does it hurt the bee more?
Do bees die after stinging?

FACT FILE

Honeybees use a special movement known as a waggle dance to tell each other where to find the best nectar and pollen. The waggle dance shows how far away the flowers are and in which direction.

DO ALL BEES LIVE IN HIVES?

No. Only social bees, like honeybees and bumblebees, live in hives. Most bees live alone or in small groups in other places, such as underground.

DO ALL BEES MAKE HONEY?

No. In fact, less than 5% of the bee species in the world make honey. The best-known bee is the honeybee, which also produces beeswax.

PARTIALLY TRUE

It depends on the species. Female honeybees have barbs on their stingers. If a honeybee stings a creature with thick skin (like a human), the barbs get stuck in the skin. As the bee tries to fly away, part of the bee's body is ripped away. The bee will soon die from the injury.

SOMETHING SMELLS SNAKE-Y

A snake often flicks its tongue in and out.
Does a snake use its tongue to smell?

FACT CHECKERS SAY...

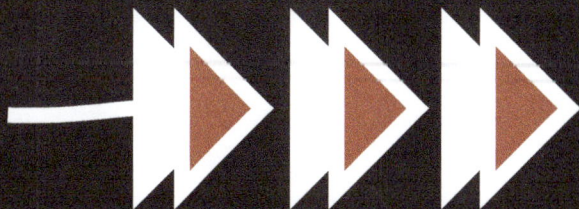

FACT FILE

The Eastern Coral Snake is sometimes called the harlequin snake. A harlequin is a theater character who often dresses in colorful clothing.

PARTIALLY TRUE

A snake doesn't use its tongue to smell things, but its tongue helps in the process. When a snake flicks its tongue, it picks up scent clues from the air and brings them to the Jacobson's organ on the roof of its mouth. This organ then figures out what the scents are.

FACT CHECKER REPORT

DO VENOMOUS SNAKES HAVE TRIANGLE-SHAPED HEADS?

Not necessarily. Many venomous snakes do have triangle-shaped heads, but some non-venomous snakes flatten their heads when threatened. This behavior makes their heads appear larger and triangular.

DO VENOMOUS SNAKES HAVE VERTICAL PUPILS?

Some venomous snakes have vertical pupils, but others have round pupils. This pit viper has vertical pupils. The coral snake (opposite) has round pupils. Both snakes are venomous. Pupil shape does not determine whether a snake is venomous or non-venomous.

DOES A SNAKE UNHINGE ITS JAW TO EAT?

While it might look like a snake dislocates its jaw to eat, that isn't the case. A snake has more than one hinge point because of the extra bones in its jaw. Its mouth also has many tendons so it can stretch side to side and up and down.

WATER LOGGED
Does a camel store water in its hump?

NOT TRUE
A camel stores fat in its hump(s), not water. The camel uses this fat for energy when food is scarce.

SANDY SITUATION

Does an ostrich bury its head in the sand when it senses trouble?

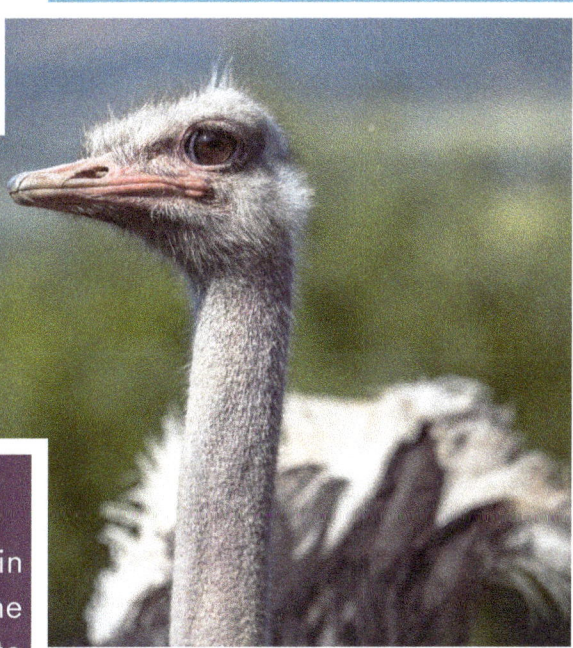

NO

An ostrich doesn't bury its head in the sand. It does make its nest in the sand, and it lowers its head often to turn its eggs in the nest. This may give the appearance that it buries its head in the sand.

FACT FILE

When it senses danger, an ostrich may stretch out flat on the ground to avoid being noticed. This might also explain why people think an ostrich buries its head in the sand.

FACT CHECKER REPORT

DO CAMELS SPIT AT PEOPLE A LOT?

Camels spit when agitated. They spit up not just saliva, but also whatever is in their stomachs.

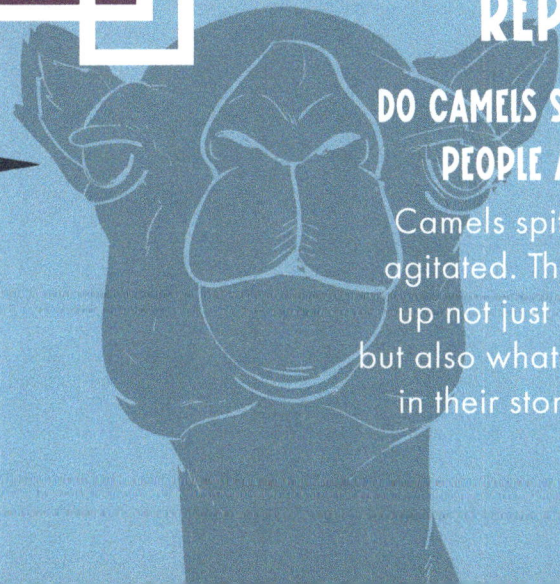

FACT & OPINION FILE

There are around 400 dog breeds in the world. Which dog breed is your favorite?

SHADES OF GRAY

Do dogs only see the world in black and white?

ALMOST TRUE

Dogs don't see the same number of colors as humans do, but they can see some colors. Their sight may be similar to that of a person who is colorblind.

CAN YOU TEACH AN OLD DOG NEW TRICKS?

Old dogs learn more slowly than young dogs, but you can teach an old dog new tricks.

IS ONE HUMAN YEAR EQUAL TO SEVEN DOG YEARS?

Dogs age faster than people, but there isn't an exact correlation between human years and dog years.

SUPER SNIFFER

Sharks have a blood-thirsty reputation. Can sharks really smell a single drop of blood in the water from a mile away?

FACT CHECKERS SAY...

NOPE!

Sharks can smell blood in the water, but not from a mile away. Scientists think it is more likely that sharks can smell a drop of blood in a body of water the size of a swimming pool.

FACT FILE

The whale shark is the largest fish on Earth. It can grow to as much as 40 to 60 feet long! However, it is harmless to people. It eats mostly plankton.

FACT CHECKER REPORT

DO SHARKS HAVE BONES?

Sharks do not have bones. Their skeletons are made of strong, flexible connective tissue called cartilage. This is the same tissue found in human ears and noses (and other parts of the body).

ARE SHARKS BLOOD-THIRSTY ANIMALS?

Like humans, sharks are predators. However, this does not mean they are blood-thirsty. This myth probably comes from the stories about sharks attacking humans, but these attacks are very rare. You have a better chance of being hit by lightning!

DO SHARKS NEED TO SWIM CONSTANTLY TO STAY ALIVE?

Some sharks must swim to keep water moving through their gills. The great white shark and the whale shark are two examples. However, sharks like the tiger shark and the nurse shark can pump water through their gills while they are resting.

SEEING RED

Bullfighters are known for the bright red capes they wave in front of bulls. The bulls then charge at the cape and the bullfighter. Are bulls enraged by the color red?

FACT CHECKERS SAY...

TOUGH TO SWALLOW

Goats have a reputation for eating everything. Do goats eat tin cans and other kinds of garbage?

FACT FILE

Like most grazing animals, goats have pupils that look like horizontal rectangles. This helps them watch for predators while they graze.

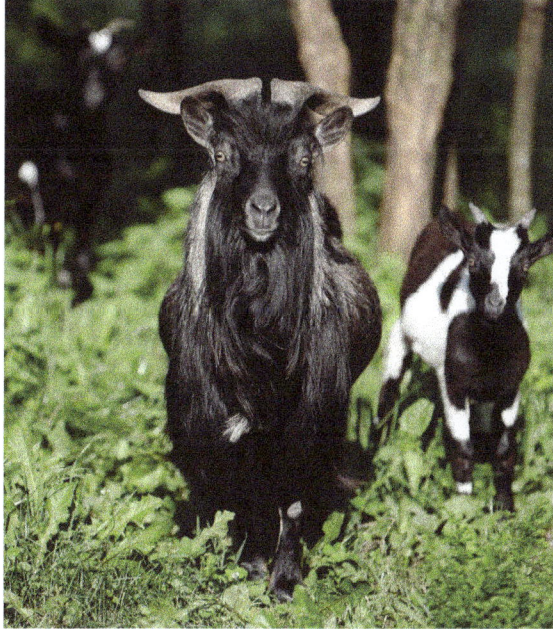

NOPE!

Goats are curious. They will explore anything in search of food—even a tin can! However, they do not eat the can itself. Goats eat roughage such as grass, weeds, and hay.

FACT CHECKER REPORT

DO COWS HAVE FOUR STOMACHS?

Not exactly.
Cows have one stomach with four separate compartments. Each compartment plays a special role in the digestion process.

NO!

Bulls dislike the movement of the cape, not its color. They get just as angry at someone waving a green and blue polka dot flag!

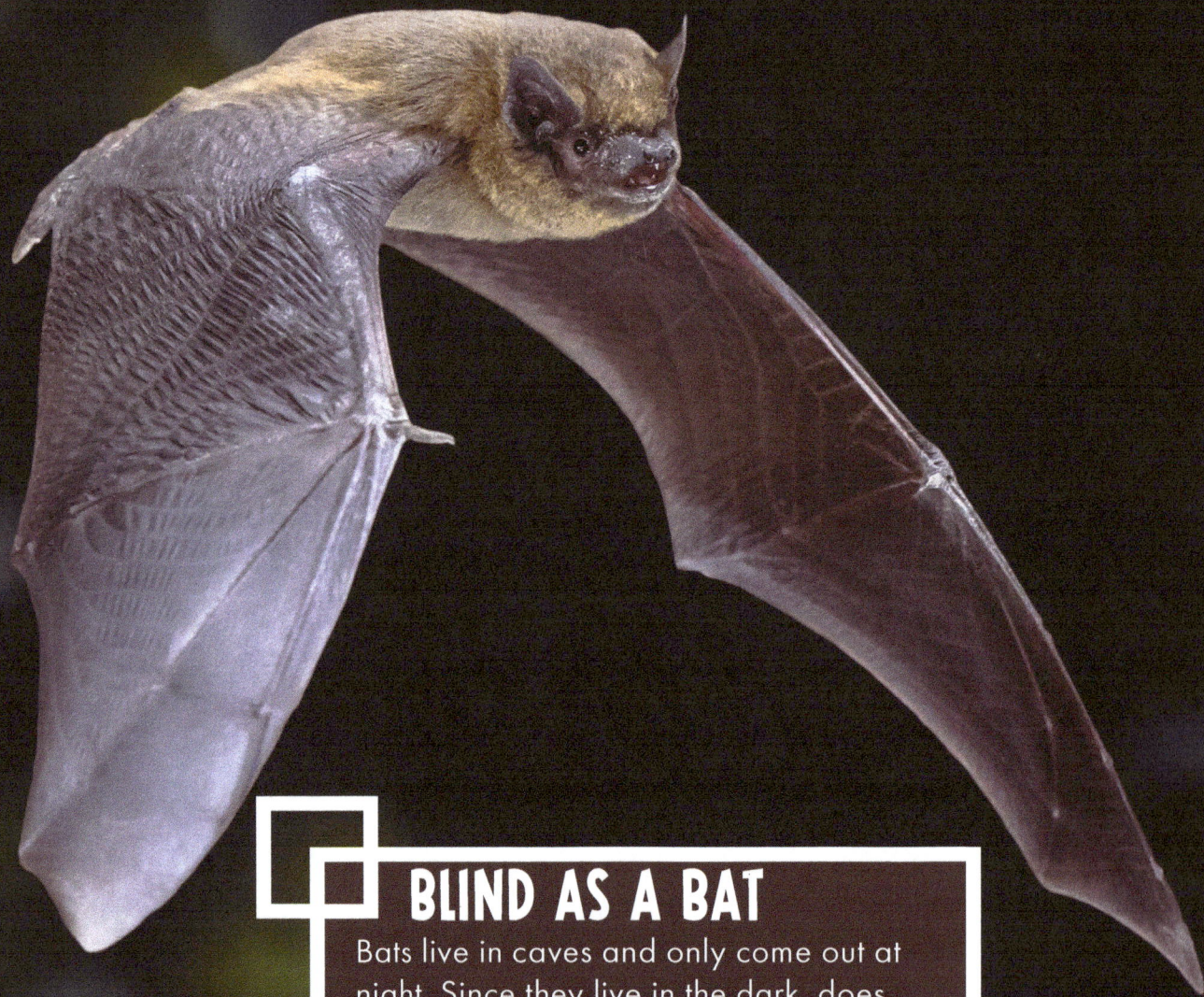

BLIND AS A BAT

Bats live in caves and only come out at night. Since they live in the dark, does that mean bats are blind?

FACT FILE

The flying fox is the largest bat in the world. Its wings can span up to five feet!

FACT CHECKER REPORT

DO ALL BATS DRINK BLOOD?

Not all bats. Three species of vampire bats drink the blood of other mammals. They do not suck the blood like people believe. Instead, they lap it up from a small bite they make with their sharp teeth.

ARE ALL BATS RABID?

Some bats do carry rabies, but most bats do not. Bats can pass rabies to humans. You can't tell if a bat has rabies by looking at it. It is best to assume any bat you see has the disease, especially if it seems sick.

NOPE!

People probably believe this because bats are active mostly at night and have great hearing. It's also true that bats often rely on echolocation to "see".

GLOSSARY

Definitions for selected words found in this book.

AGITATED (adjective) anxious, restless, or upset

BREED (noun) a group of animals with similar characteristics

CHEMICAL (noun) a substance with unique properties that is made of one or more elements

COLORBLIND (adjective) difficulty seeing certain colors or differences between colors

CONVERT (verb) to change something into a different form or use

CORRELATION (noun) a relationship between two things where changes in one of the things might cause changes in the other

DISLOCATE (verb) to move a bone out of its normal position in a joint

ECHOLOCATION (noun) using sound waves to find objects and navigate

ENRAGED (adjective) very angry or furious

GILL (noun) a part of a fish and some other aquatic animals that helps them breathe in water

HERBIVORE (noun) an animal that eats only plants

HIVE (noun) a structure where bees live, work, and store honey

HORIZONTAL (adjective) straight across

MOLAR (noun) a wide, flat tooth in the back of the mouth used for chewing

MYTH (noun) a false belief

NECTAR (noun) a sweet liquid found in flowers that many insects and birds use for food

NYMPH (noun) a stage in the life cycle of some insects during which the young look like small adults but without wings

PLANKTON (noun) tiny organisms that float in water that many other animals eat

PUPIL (noun) the opening in the center of the eye that lets light in

PREDATOR (noun) an animal that hunts and eats other animals for food

RABID (adjective) having rabies, which is a dangerous disease passed through the bite of an infected animal

REPUTATION (noun) the opinion people have about someone or something, based on behavior or character

SCARCE (adjective) in short supply

SILK (noun) a soft, shiny thread some spiders use to make their webs

SPAN (noun) the distance between two things

SPECIES (noun) a group of living organisms that are similar and can reproduce with each other

TENDON (noun) a strong tissue that connects muscles to bones

VENOMOUS (adjective) full of venom, a substance some animals produce that can be harmful to animals or people

VERTICAL (adjective) straight up and down

Text copyright © 2025 Kizzi Roberts and Carrie Rodell

Photographs © imagex/depositphotos.com; PantherMediaSeller/depositphotos.com; izanbar/depositphotos.com; JohanSwanepoel/depositphotos.com; iwayansumatika/depositphotos.com; Ondreicka1010/depositphotos.com; REPTILES4ALL/depositphotos.com; cookelma/depositphotos.com; Vlad61/depositphotos.com; alexeys/depositphotos.com; peternile/depositphotos.com; frenky362/depositphotos.com; Tomatito/depositphotos.com; graphicphoto/depositphotos.com; imagebrokermicrostock/depositphotos.com; GarryKillian_/depositphotos.com; Curioso_Travel_Photography/depositphotos.com; paulmaguire/depositphotos.com; NaDo_Krasivo/depositphotos.com; CreativeNature/depositphotos.com; Wirestock/depositphotos.com; Ksuksann/depositphotos.com; mythja/depositphotos.com; albertoclemares.hotmail.com/depositphotos.com; stokkete/depositphotos.com; frenta/depositphotos.com; photolime/envato.

All rights reserved. No part of this book may be reproduced or used in any manner without the prior written permission of the copyright owner, except for the use of brief quotations in a book review. To request permissions, contact the publisher at permissions@LearningSpark.com.

Published in May 2025 by Learning Spark Educational Publishing in Rogersville, Missouri. Learning Spark Educational Publishing is an imprint of Elemental Ink Publishing LLC.

Library of Congress Control Number: 2025905278

Hardcover: 9798888840337; Paperback: 9798888840221; Ebook: 9798888840344
Edited by Carrie Rodell. Book design and layout by Kizzi Roberts.

www.LearningSpark.com

www.ingramcontent.com/pod-product-compliance
Lightning Source LLC
Chambersburg PA
CBHW041600260326
41914CB00011B/1334